Preface

There are many belief systems in the world; some are very similar to each other and some very different. Most, if not all, believe in a creator. A supreme all-powerful, all-knowing being, that bought the universe into existence.

Most traditional religions believe that the universe, and our place and role in it, was brought into existence by a supreme being for a purpose. There is a reason that we exist. There is a reason why the universe exists. There is a reason why things are the way they are. It is not a matter of simple coincidence or chance.

But science, and most scientists, have a different view on how the universe came to be. Scientists do not rely on the existence of a deity to account for us and the universe. Scientific discoveries over the centuries have led them to believe that the universe can be described using mathematical equations. It's all a matter of physics and chance. And more and more people are turning to this atheistic view of the universe.

But it's possible atheism itself may be a form of religion. This statement will probably be anathema to scientists, cosmologists, and physicists. This is because religion is about faith and belief. Belief in something you cannot see or show evidence for. Whereas science is all about evidence and proof.

Faith has nothing to do with a scientific theory. If a physicist comes up with a new hypothesis or a new theory on the working of our universe it will be published, peer reviewed, pored over by countless other scientists to consider its validity. Experiments will be conducted and if evidence is uncovered to show it matches the theoretical predictions, then scientists will start to take it seriously. And if over time no holes can be found to debunk the theory, it will become part of standard scientific knowledge.

Where the data leads; scientists will follow. If the data disproves the theory, it will be abandoned. If it indicates only certain aspects of it are evidenced, then it will be modified.

This is clearly the way things should be.

It has been very easy in the past, and unfortunately with the advent of the internet and social media, in our current times as well, to make up some conspiracy theory, for example, make it go viral and have people believe in it without understanding or querying the source and whether it is true or not. Or even if there is

A CHORUS OF BIG BANGS

Adam Susskind

any evidence for it to be true. Truth and facts are harder and harder to decipher from all the information now bombarded at us from so many directions.

So, I have a lot of admiration for physicist, cosmologists, astronomers, astrophysicists, whom I collectively call scientists in this book. I respect all these seekers of knowledge and truth. So many of whom dedicate their whole lives carrying out research, conducting experiments and coming up with theories and predictions that expand the knowledge of humanity. They are all heroes.

The problem is that scientists are meant to have an open mind. And generally, they do. They send massive telescopes out into space, build tunnels underground that cost billions to look for new particles, they pore over the data and follow where it leads.

But there is one area in which it seems their minds are not as open as they could be. An area where the data and the evidence are leading them somewhere they don't really want to go. And they find ways not to go there despite the evidence being right in front of them.

And the evidence that they have uncovered is that there is a very strong possibility that the universe had a creator. It simply cannot have come about by chance. The odds that the universe came into being just by chance are so astronomical that there are more zeroes in the odds than the human mind can comprehend.

This is not a statement I am making based on any new theory I have come up with myself or any new data that I have uncovered. This is a fact accepted by the scientific community based on evidence accumulated over decades. Ask any scientist and they will accept that the chance that the universe, as it exists and allows us human beings to exist, came about just by chance are just mind-numbingly long.

However, most scientists simply cannot accept that the universe has a creator. If the universe was created by someone then what would be the point of science? Miracles cannot be tested or proven. They are of divine origin. A universe created by a miracle leaves science redundant. This is not necessarily true, but it does seem to be the prevailing view. Consequently, the possibility that the universe was created is not something most scientists are willing to consider.

I am reminded of a Netflix documentary about people who believe that the Earth is flat called "Behind the Curve". In one segment a few of them carry out an experiment to prove that the Earth is flat. The experiment fails, and in fact the experiment proves that the Earth does have the curvature expected and consequently cannot be flat. But being strong believers in the flat Earth theory they simply cannot accept the evidence. So, the response is well, we have other experiments we can try that will prove the Earth is flat.

Scientists may scoff at being compared to flat Earthers, but the analogy is apt. Most scientists do not believe in a creator. They cannot believe in a creator. But they have all this evidence to show that the universe cannot have come about by chance. And it's just too overwhelming to ignore.

So, what do they do in the face of this evidence?

What they have decided to do is to take a leap of faith that is as big as any taken by someone who believes in the existence of a creator and come up with a theory for which there is not only no evidence, but no way of even getting evidence. This theory should be scientific heresy since it cannot be tested or proven. Ever.

But the theory does allow for an elegant way to bypass all the evidence science has uncovered that there is at least the possibility that the universe had a creator.

The aim of this book is two-fold. One is to highlight that scientists, like other believers in religion, are in some sense also people of faith. And that their religion could be classed Atheism.

The second aim of the book is to highlight to regular readers who are not scientists the gaps in the scientific theories, that scientists freely admit exist, but of which readers may not be aware.

And also, to point to the evidence that scientists have uncovered that strongly indicates that the universe cannot have come about by chance. I hope that it will give them food for thought.

The evidence I have gathered for this book has been from freely available articles, journals, documentaries, lectures and one-to-one interviews of respected, well known, and not so well known, astronomers, physicists, and cosmologists. I have quoted liberally from these individuals so that the story can be told in their own words to avoid personal bias.

I have attempted to be as balanced as possible and not cherry pick specifics, or use out of context quotes, to allow the reader to come to a fair conclusion on the premise of this book.

And lastly it must be noted that this book is not designed to be a technical work. It does not have diagrams, or graphs or any equations. It is written in very simple narrative terms from the point of view that any individual with or without a scientific background can follow.

PART ONE - CREATION

The Greatest Story Ever Told

In the beginning was a Big Bang.
The Big Bang begat a cloud of gas.
The stars grouped into galaxies.
And the Universe came to be

Amen

Chapter 1

Like all true believers, cosmologists also have scriptures. Theirs are entitled "The Standard Model of Cosmology" and "The Standard Model of Particle Physics". Unlike the Torah, the Bible, or the Quran these scientific texts were written quite recently. Like the bible though, they have many authors.

Scientists would argue that, unlike the traditional holy scriptures, theirs are based on research, experiments, evidence, and facts. They don't rely on a supreme being creating the universe and everything in it out of nothing. They have evidence that it was all really very simple. It all began with a big bang.

Martin Rees - the Astronomer Royal

"The Big Bang in a sense was rather simple. You can write down a rather short recipe. From that, in principle, if you had a powerful enough computer, work out what would happen and end up with something rather like our universe."

So, what is the short recipe?

Max Tegmark – Professor MIT

"Traditionally we use physics to start with the present to predict the future. You can also do it backwards. And

start with what we see around us in the present day and calculate what the universe must have been like in the past in order to end up like this. And that is exactly how we have come to the conclusion that there was a Big Bang."

And that makes sense. Scientists have verified that the universe is expanding. Getting bigger and bigger every day. So logically, if you work backwards the universe will get smaller and smaller and you can imagine a point where it first began.

Since light has a finite speed, scientists have been able to verify this to a certain extent. This is because light from the farthest stars can take millions and billions of years to reach planet Earth as we are so far away.

In effect when scientists point their telescopes to the skies and beyond they are actually looking back in time. As telescopes become more and more sophisticated scientists can look further and further back in time.

Using this precept scientists have taken the physical events we see around us , like how big is the universe, how many galaxies there are, how our solar system works, and turned them into mathematical equations. And the equations changed and moved around to understand how things worked in the past. This in turn enabled scientists to create a mathematical replica of the universe.

And so, the first bible of the Scientific world, "The

Standard Model of Cosmology" came into being.

Carlos Frenk – Prof. University of Durham

"The Standard Model of Cosmology describes the entire evolution of the cosmos from an instant after the Big Bang to the present day."

The Big Bang theory is quite simple. It goes something like this:

About 13.8 billion years ago there was nothing. Absolutely nothing. Even time and space did not exist.

Then out of that nothing there was suddenly something. A big bang.

After just one second of the bang, energy was turned into matter filling everything with a dense fog.

After an eon of time the fog settled into the first stars by way of gravity pulling matter together.

The stars formed into galaxies.

And galaxies led to planetary solar systems.

And then eventually to us.

All very simple and as predicted by mathematical models.

Alan Guth – winner of the Franklin Medal for Physics

"It's a remarkable theory as it really does allow us

to understand literally everything we can see. It works perfectly."

Well, not quite perfectly. There are a few holes that scientists have discovered that don't allow the first bible of the Scientific world to quite add up.

Scientists readily admit that the holes in the model are rather big.

Chapter 2

The very first, and possibly the biggest problem, is the big bang itself.

Where did it come from?

Cause and effect. This is one of the key scientific principles. This thing happened because of this other thing which in turn led to this other, other thing. This is in fact how scientists have pieced together the Standard Model of Cosmology. From where we are currently back to the billionth and trillionth and quadrillionth of a second after the big bang.

But what caused the big bang to happen in the first place?

There was no time, no space. Nothing. From out of that nothing there was suddenly something. A bang. A big bang even. But where did the big bang come from?

Some scientists believe this is not a question that makes any sense.

Metin Tolan- Experimental physicist

"This is a question about the origin of everything.... Maybe our brain isn't even equipped for dealing with questions like this. The big bang is the origin of everything. When you ask what came before the big bang, a physicist would

almost have to say that that's a question that makes no sense because time itself only came into being with the big bang. It's impossible for anything to have existed before it. Some questions in science make no sense."

Metin is saying that time began with the big bang so how can there be a before. There was no "before" the big bang. For something to cause the big bang to happen there has to be a before, but time only started with the big bang so there is no "before". Thus, there is no cause.

It just happened.

Basically, scientists just don't know. They feel they can measure what happened a trillionth of a second after the big bang. But they don't know what caused it to happen. It's not a question scientists can answer.

When scientists are stumped and don't know something they like to give the thing they don't know a name. In the case of what caused the big bang to happen they have named it a "Singularity."

The big bang came out of a "Singularity."

A Singularity is a place in the universe where the laws of physics simply break down. The laws of nature that scientists have managed to decipher just don't work in a singularity. It's a place where no known laws of nature apply. It's an anomaly. A mystery wrapped around a riddle.

Max Tegmark – Professor MIT

"What's so embarrassing about a singularity is that we can't predict anything about what's going to come out of it. I have a singularity here and boom out comes a pink elephant with purple stripes. And that's consistent with what the laws of physics predicts. Because they don't predict anything."

Since, it is possible for a singularity to produce a pink elephant with purple stripes from out of nowhere, then there is nothing to stop it from say creating the universe in six days out of nothing if it really wanted to.

But from a scientific perspective when you give something that you don't know a name then you can almost act as if the question has been answered. Science has given what caused the big bang a name so can ignore it and not think about it too much for now. We know this is an issue, but we will figure something out. It's not a big deal. We're working on it.

Move on, there is nothing to see here.

But there is something to see here. Something very fundamental. If you are going to close your mind to the possibility of a creator, then surely you should have a legitimate explanation for what made the big bang happen. How can the universe come into being just by itself?

To be fair, scientists have come up with a few wild and exotic theories about what caused the big bang to happen.

Roger Penrose - mathematical physicist, and Nobel Laureate

"The present picture of the universe is that it starts with a big bang, and it ends with an indefinitely expanding universe. And in the remote future there's not much left except photons.....and the universe loses track of how big it is. And this expanded universe becomes equivalent of a big bang of another one. What we think of our present universe is but one eon of a succession of eons, where the expanded universe becomes the big bang of the next."

According to this hypothesis, as one universe collapses it kicks off the big bang to begin the next universe. And it continues like this forever.

There is another theory that proposes that the big bang came out of a black hole.

Lee Smolin – Prof. theoretical physicist

"Before the big bang there was another universe much like our own. In that universe there was a big cloud of gas and dust. It collapsed to form a big, massive star. That star exploded, it left behind a black hole, and in that black hole there is a region that gets denser and denser...and all of a sudden it would explode again. And that would be our big

bang."

There are plenty of such other theories.

But none of them can provide cause and effect. They just delay the cause. And there is no evidence or proof for any of these theories anyway. It's mostly just speculation.

It's clear that scientists just don't know why the universe came into existence.

At this point you would think scientists with open minds would contemplate that well, maybe, just maybe, there is a slight possibility that the universe was created. That it did not just come into being out of nothing.

But this is not a road most scientists want to go down. Sooner or later science will find an answer.

But what caused the big bang to happen was just the start of the holes in the cosmological bible that scientists were starting to uncover.

Chapter 3

I love balloons. They are fun and provide a party atmosphere. And they come in lots of fun, bright colours. When you buy a packet of balloons you take one out and start blowing it up. Of course, you are careful not to overdo it and stop inflating it just before the point it would pop, and you are left with a crying child.

It seems balloons and the big bang theory have something in common. They both require inflation, but up to a very specific point.

The Standard Model of Cosmology has been built around several theories that explains how the universe came to be in its current state. But scientists have found some flaws in the model. And these flaws are rather big.

The first flaw is why the temperature of the universe is almost exactly the same everywhere across the universe. If the universe was created by a big bang, then the universe as we see it today should be lumpy and uneven. But it isn't.

The original big bang theory could not account for the uniformity.

Alan Guth – winner the Franklin Medal for Physics

"There were things just left out of the big bang theory. There is no understanding of the uniformity we see in the universe. The universe looks essentially the same if we look that way or that way. No normal explosion would ever do that."

But Alan came up with an answer. The theory of inflation.

"Inflation gets around this problem essentially by varying the expansion history of the universe. So that the universe starts out essentially dawdling with low expansion rate for a period of time before inflation.

And it's during that time that the universe is extremely tiny and not expanding that rapidly that it can come to a uniform temperature. Then inflation takes over and magnifies that tiny region to become large enough to include everything we see and likely far beyond"

Problem solved.

The universe started very small; it remained like that for a period of time allowing the temperature to be the same everywhere. And then it expanded massively in an instant which left everything perfectly smooth and uniform.

Although it was just a theory scientists managed to discover some compelling evidence to not necessarily prove but at least support the theory.

Using satellites, cosmologists photographed light from the very early universe. Since light travels at a finite speed, they were able to look back in time some 380,000 years after the big bang.

The picture they built up showed a special type of radiation called the Cosmic Microwave Background, and the CMB showed that the temperature was the same in all directions. This gave some credence to the inflation theory that at least 380,000 years after the big bang the universe had the same uniformity as we see now. So somewhere in between the first second or two after the big bang and 380,000 years later the universe went from lumpy to smooth and uniform.

However, the theory has some very serious flaws.

We again have to go back to cause and effect. What caused inflation to happen?

Pedro Ferreira – Cosmologist

"We don't know how inflation happened. We don't know what drove the universe to inflate."

Inflation would have needed a force so powerful that is has never been seen before.

Max Tegmark – Professor MIT

"Right now, we are nowhere near being able to create enough energy in our laboratories to create inflation or

anything like it."

Another problem with inflation is that, like being able to stop inflating a balloon at the just the right time that it doesn't pop, how did the universe suddenly stop inflating at just the right time before the universe blew apart?

Some scientists have strong doubts about the inflation theory as it just seems too contrived.

Roger Penrose - mathematical physicist and Nobel Laureate

"I rather disagree about inflation being a nice, beautiful theory. It's a very artificial theory, where you have to introduce concepts that are invented specifically for the theory, they don't come from anything else."

This quote from Dr Kathy Romer - University of Sussex is rather interesting:

"I'm not a great fan of inflation.... I don't like it because it is too much of an add on. But saying that, it is the easiest way for us to get from the standard model to some unexplained observations."

There is clearly some cognitive dissonance going on here.

Cosmologists and even Alan Guth, the proponent of the theory, accept that the theory of inflation is seriously flawed since there is no cause and effect. It just happened. And then stopped happening at exactly the right time so the universe wouldn't blow apart. And apart from the CMB picture there is no other evidence to support the theory.

It really is just an invention. But the invention keeps the belief in the standard model intact. So as seriously flawed as it is, it cannot be thrown away and has become a key part of the standard model.

But if having to shoehorn in the theory of inflation to make the standard model work wasn't bad enough, things got even more darker and trickier for scientists to keep the faith.

Chapter 4

Nobody likes going up hills. It takes so much effort. And it's so tiring. Is it really worth the pain of getting up there to see the view? By the time you get up there you are so out of breath that you don't enjoy the view anyway!

Going down is much more fun. It's so much easier! When riding a bike going down no pedalling is required. Just let the force of gravity do its work.

Gravity.

So simple and yet so complicated. It keeps everything bound together in a stunning, amazing, beautifully mathematical way. It keeps our feet on the floor, our planet circling our sun, our solar system circling our galaxy and our galaxy circling our universe.

It's a beautiful thing. And we need it. Without gravity everything would just fly away from each other. And we simply would not exist.

But for the standard model gravity is a big problem. Quite a massive problem.

Our solar system consists of eight planets. Mercury is the closest, then Venus, then Earth, then Mars. These are known as the inner planets as they are close to the sun. Then you get the outer planets which are much bigger and much further away from the sun, Jupiter, Saturn, Uranus, and Neptune.

Gravity in our solar system works exactly as predicted by all the theories. The closer a planet is to the sun, the stronger is the force of gravity it exerts over the planet and the faster it orbits the sun. The further away a planet is the weaker is the force exerted and the slower it rotates around the sun.

Stars also orbit their galaxies, and they should behave in the same way as the planets do in our solar system. The further away the stars are from the centre of the galaxy the slower should be their rotation speed.

But in 1974 astronomers discovered that that is not the behaviour seen.

Bob Nichol – Astrophysicist – University of Portsmouth

"In the solar system we have the sun in the middle which provides all the gravity. And then coming further out from that we have all the planets that are lined up and rotate around the sun. And the speed by which they go round the sun decreases as a function of the distance away from the sun.

So, by the time you get to the outer planets they are moving a lot slower than the ones in the centre. So, Neptune (the planet furthest away from the sun) takes 165 Earth years to go round the sun...

We have the same set up in our galaxy. We have a large super massive black hole in the centre, and we have stars orbiting around the centre of the galaxy. So, you would expect the stars further away from the centre of the galaxy would be moving slower than the ones on the inside.

But that's not what we see. What we see is the speed of the stars is constant with distance. The stars out here (at the edge of the galaxy) are traveling at the same speed as the stars in the centre."

And wherever the speed of stars was measured by astronomers in spiral galaxies they showed the same incomprehensible flat rotation curves. The stars at the edge of the galaxies are moving just as fast as the stars close to the centre of the galaxy.

At the speeds they were moving the universe should just fly apart. But something, some glue was holding it together. But what was it?

It was inexplicable. There simply was not enough matter that could be seen in the universe to stop the universe from flying apart. It only made sense if there was more matter in the universe creating more gravity, allowing the stars further away from the centre to rotate as fast as the ones closer in.

But where was it? All the atoms and molecules and particles that scientists were aware of in our universe just could not account for this extraordinary behaviour of the star movement.

As we are aware when scientists don't know something they love to give the thing they don't know a name.

Carlos Frenk – Prof. University of Durham

"When we look at the way things move in the cosmos, we soon reach a very profound conclusion. There isn't enough gravity in the stuff we can see, the stars and galaxies, to explain how objects move in the universe. There must be something else that is responsible for this movements. And that is what we call Dark Matter."

And the name Dark Matter was born. A pure invention to shoehorn in to be able to account for the fact that without it the Standard Model of Cosmology just doesn't work.

João Magueijo– Theoretical physicist

"Dark matter is the easiest way to explain why the galaxies work the way they do and everything else."

More shocking still was how much extra of this unseen matter was needed to match the behaviour seen of flat rotation curves of the stars.

Pedro Ferreira – Cosmologist

"We expect that for every kilogram of normal matter there is another five kilograms of dark matter. And we expect that dark matter to be everywhere clustered around us."

That means there is 500% more of this invented, unseen, dark matter, than there is of the normal matter that can be accounted for.

The universe is evidently made up of matter that is completely alien to science and we know nothing about it. But without it the scientific cosmic bible just does not fit together logically.

Carlos Frenk – Prof. University of Durham

"Dark matter in general I think is one of the pillars of the Standard Model of Cosmology. It's needed. There is no doubt about that."

But what is dark matter? It is not ordinary matter as ordinary matter either emits light or reflects it. Dark matter is totally invisible and cannot be detected by any kind of telescope, optical or otherwise.

It must be a special type of matter made of a particle or particles never seen before.

João Magueijo– Theoretical physicist

"What's even worse is not that it's just invisible matter but is basically exotic matter. Things which are not atoms, the things we know."

This dark matter particle cannot be accounted for even in the second scientific bible, "The Standard Model of Particle Physics."

Carlos Frenk – Prof. University of Durham

"Today we have The Standard Model of Particle Physics in which there are twenty-four elementary particles. And they have their own values and charge and spin, and they interact with each other in different ways...none of the particles would be a candidate for dark matter."

Scientists have tortured themselves for years and decades since its discovery inventing new and exotic particles to account for dark matter. Particles with the names of Axions and Hooperons and such.

They have hunted for them doing experiments in underground mines and shafts. Desperately trying to get time on the world's best telescopes situated at high altitude in exotic locations. Pleading with the authorities at CERN to conduct experiments in the LHC, the largest underground particle accelerator ever built, to find proof for their pet theories.

And even sending out telescopes in space to gather more information on what dark matter could be.

Scientists have formulated ideas about cold dark matter and hot dark matter and theories called super symmetry which doubles the number of particles in the current standard model of particle physics to account for all the dark matter.

So much time, money and effort has been expended to

get to the bottom of the so-called dark matter.

David Charlton – Prof. ATLAS Collaboration Spokesperson, CERN

"As an experimentalist it's really my job to have an open mind and really to look at all of the possibilities and try and explore everything we might discover. The theorists might have their own favourite theories and say you should discover Super Symmetry, or you should discover something else I don't know. Nature will tell us what's there."

But so far nature is keeping her own council.

Some scientists are beginning to wonder if there even is such a thing as Dark Matter.

João Magueijo– Theoretical physicist

"Maybe it's not like that. Maybe we should just take things at face value and what's out there is what there is. There's no dark matter or ghosts around the universe."

But cosmology needs dark matter, without it the first and even second bible of the scientific world has another massive hole that it can't account for. But faith is faith. And since scientists have a name for this thing that they can't explain the standard model bible remains intact.

But if the absence of evidence of the invented dark matter was bad news, then a new problem scientists discovered in the late 1990s took things to a whole new level with trying to keep faith in the credibility of the model.

Chapter 5

Energy. You could do so much if you just had the energy to do it. Go out running, exploring, mountaineering, kayaking, ski jumping. All manner of possibilities. There is so much you could do. And it all sounds so much fun.

If you just had the energy to do it all.

We may not have energy to do all the things we want to do but it turns out the universe has much more energy, and is much more dynamic, than the Standard Model of Cosmology ever predicted.

Much more. Scientists were shocked to discover how much more.

In 1998 Saul Perlmutter was looking to carry out some meaningful research as a graduate student.

Saul Perlmutter – Lawrence Berkeley National Laboratory

"I really wanted to find a project that would answer some, or at least looking at some, very philosophical questions, something that felt meaningful of the world we live in, in some deep way.....like, whether the universe will last forever, do we live in a universe that is infinite or some day will come to an end.

The two big options at the time were that the universe could expand forever, but slow and slow, but forever be expanding or if there was enough stuff in the universe to gravitationally attract it, it could slow to a halt and then collapse and come to an end."

If the Big Bang Theory was looking back in time, then Saul was doing the opposite and looking into the future to examine whether the universe would continue to expand slowly forever or come to a stop at some point and collapse.

Those were the only two options that the Standard Model of Cosmology predicted.

The results of the research uncovered something completely and totally unexpected. So wholly unexpected that it took scientists to the very edge of their faith and belief in the Standard Model of Cosmology. Maybe even pushed some over the edge.

Bob Nichol – Prof. University of Portsmouth

"Everybody knew Saul. And everybody knew the experiment he was doing. And I remember sitting in the audience and Saul getting up and expecting him to present an update on the results he had given a year ago that actually the universe was slowing down and so I was absolutely amazed....suddenly he was saying we lived in a universe that was accelerating. I remember it being, just incredible. All the astronomers walking around, scratching

their heads saying "this can't be right. Surely this can't be right.""

It's difficult to get across how much of a shock this was to scientists. The only options were that the universe was expanding at a slow rate and would continue expanding forever or that at some stage it will stop expanding. What they were not expecting the results to show was that the expansion of the universe was not slowing down but actually accelerating.

No way should the expansion of the universe be accelerating.

Scientists were just getting over the Dark Matter problem, thinking at some stage they will find a solution for it, but now this thing. This was even bigger than the Dark Matter anomaly.

Again, like all the previous anomalies uncovered, none of the theories could account for why the universe was expanding at such a rate. Not even the General Theory of Relativity or Quantum Mechanics, the two theories that although were contradictory and didn't work well together, were still the bedrock of scientific discovery.

Another unwanted, undesired spanner mucking up the works of the Standard Model of Cosmology idea.

A can of worms.

Of course, when something like this happens, scientists

did what they like to do. They gave the problem a name; "Dark Energy."

And they started looking at what could possibly be causing the universe expansion to be speeding up. And the more they learned about it the more mystifying it turned out to be.

It seemed like the more the universe expanded the more new "Dark Energy" was being created. This just should not happen.

Bob Nichol – Prof. University of Portsmouth

"So, you can think of it, as you get more space, you actually get more dark energy. Which is like getting something for nothing. Which is clearly ridiculous. I mean it's clearly against all our training as physicists."

What was worst was how much extra energy was required to meet the observations of a rapidly expanding universe. It was a lot.

The discovery of dark matter and now dark energy brought into question what the universe was even made of!

Physicists had thought they had figured that out already. Atoms. Everything was supposed to have been made from atoms. The stars, the planets, the moons, bowling balls, tennis rackets, chocolate puddings and even cats and dogs. We and the universe are all made from atoms. Which are in turn made from protons, neutrons, and electrons.

But it now turns out that that the universe is made of more than just atoms. There is so much stuff out there somewhere that what scientists have been able to see, and measure, is only a fraction of the matter that makes up the universe.

After doing lots and lots of calculations scientists arrived at the conclusion that the universe was made up of about:

05% baryonic matter. - Scientists know about.
27% dark matter - Scientists don't know about.
68% dark energy - Scientists don't know about.

This means that the first bible of the scientific world cannot account for 95% of what the universe is made of.

That is 95%.

95%.

Dr Clare Burrage - University of Nottingham

"The fact that our predictions are so far off from what we see tells us that there is something fundamentally missing in the way we understand physics and the way we understand the world."

Unwanted, and unloved though it may have been,

scientists are always looking for holes in their theories. And are quite pleased about it when they find one because it means they have uncovered more knowledge and a greater understanding of the workings of the universe. And that is good thing.

The hole that Saul uncovered was so big that he even got a Nobel prize for his research.

Which is all great, but where does that leave the standard model?

Fully intact it seems.

Despite uncovering the fact that there are massive holes in the Standard Model of Cosmology such as what caused inflation and stopped it at just the right time, what is dark matter and dark energy, and especially the crucial question of what caused the big bang to happen in the first place, the faith in the model remains unwavering.

But how do scientists account for the gaps?

Neil deGrasse Tyson – Astrophysicist and writer

"The universe is under no obligation to make sense to you."

Interesting quote.

Here is another interesting one:

Martin Rees - the Astronomer Royal

"There's no particular reason why our human brains should have evolved just far enough to be able to assimilate the deepest level of reality."

This quote sounds rather familiar.

Scientists still believe in their bible, but it seems that some things are just beyond our understanding. The Singularity it seems has its wonders to perform, the reasons of which may never be clear to us.

This is fine. But if scientists are so keen to point out that traditional holy books don't tally with observations and evidence they have discovered in respect to creation, then they also must accept that the bible they believe in also doesn't tally in that respect either.

If certain things in your holy book are beyond understanding and cannot be addressed but you still believe in them, then surely you must be doing so as a matter of faith.

Chapter 6

There is a term that has become quite common in the scientific world called "The Crisis in Cosmology." Surprisingly this term does not refer to all the staggering gaps in the standard model as scientists have very openly noted and for which Nobel and other prizes have been handed out, like dark matter, and dark energy and the theory of inflation.

No, the crisis in cosmology is something different. The crisis refers to the fact that astronomers have worked out two different and very precise ways to measure the age and the expansion rate of the universe. And the two are both valid but they are providing slightly different results in the rate of expansion.

One way to carry out the measurement is to use the Cosmic Wave Background. This is the farthest cosmologists have been able to look back in time to the beginning of the universe, which is 380,000 years after the big bang. Then using various mathematical models and equations the age and expansion history of the universe can be worked out quite accurately.

The other way is to use exploding stars known as supernovas. Stars are very long lived, and their life spans are measured in billions of years. But eventually they do run out of fuel. And when they do, about one to three percent of them go out in a blaze of glory. Their explosions are so massive they may shine with the brightness of 10 billion suns.

These supernovas are called standard candles. And these "standard candles" explode with the same brightness. And based on how far or how close these supernovas are, scientists can work out the expansion rate of the universe.

The problem is that these two ways of measurements are providing different results. And as both calculations get more and more accurate the difference between them is getting bigger rather than smaller as scientists had been hoping.

So there is something wrong somewhere. Scientists just don't know what that something is. And of course, when scientists don't know something, they give it a name. This particular anomaly is referred to as the "Hubble Tension."

Scientists are quite perturbed by this increasing disparity. But they have given it a name, so don't have to worry too much about it.

The bigger problem though is that this, and the other more serious gaps in the Standard Model of Cosmology and Standard Model of Particle Physics, are just the tip of the iceberg.

What scientists discovered regarding how the laws of nature work just blows everything out of the water. And strongly points to the premise that, although it may not be as described in traditional holy books, the universe may have been created and did not come about by just random chance.

PART TWO - PRECISION

The Laws of Nature

In the Singularity's grand design we see,
A universe tuned for you and me.
From atoms small to galaxies vast,
Every detail, perfect and steadfast.

Amen

Chapter 1

Scientists are very transparent and are always seeking to modify current theories or seek new ones to explain the workings of the universe. Their bible is dynamic, they are more than willing to make changes to it as new discoveries are made.

And when discoveries are made that identifies gaps showing their theories are incomplete and for which they have no answer, they accept that there are things they don't know.

Yet. They don't know yet.

At some stage in the future though, solutions will be found for things their bible does not currently have answers. Some genius will come along and marry the theory of general relativity and quantum mechanics and that will explain everything. Like what caused the big bang, why inflation happened and stopped at just the right time, and what is the stuff that 95% of the universe is made up of.

They already even have a name for this theory that does not currently exist. It's called "Quantum Gravity." All they need now is a genius to come along and fill in on all the details. The important thing of naming it has already been done.

Yes sure, the bibles of cosmology and physics currently

only explains 5% of the universe, but scientists have got a pretty good handle on that 5%. They believe they know what everything that we see is made from.

Atoms.

The 5% of the universe we see, including us human beings and all life on the planet is made from atoms.

Martin Rees - the Astronomer Royal

"..We and everything else are made up of atoms...we now know that there are 92 different kinds of atoms which occur naturally. Same atoms combined in different ways make up everything. Every substance is a different combination of atoms in particular proportions."

Max Tegmark – Professor MIT

"The big difference between, on one hand, a dead thing like a crystal, and on the other hand a living thing, isn't what they are made of. Because they are made of the same building blocks, the same kind of atoms. It's rather the complexity in how they are put together.

If you want to build a squirrel, even to build one chromosome, you have to take these atoms and keep putting them in new and different ways without repeating yourself. And that's how you get this fantastic complexity which is the hallmark of life."

And the most complex object ever constructed out of atoms? That would be us.

Martin Rees - the Astronomer Royal

"We can say with some confidence that human beings are the most complicated things in the universe. The most complicated things we know about. Human beings contain ten thousand trillion, trillion atoms put together in a very complicated way."

So, the 5% that makes up everything we see and know about came from atoms. 92 different kinds of atoms.

Scientists even know where those 92 atoms came from.

It all started with the big bang of course.

In the early stages of the big bang there was just gas made from only two basic atoms, hydrogen, and helium.

But along with gas there was also a force. The force we all know and love. Gravity. The strength of the force of gravity was just right to bring the cloud of gas together make it hotter and hotter, denser, and denser, and eventually it would become a star. A star like our sun.

And the stars created the other atoms.

Steven Weinberg - Particle physicist

"Early on when the universe first began to produce stars, it started with hydrogen and helium, as the stars contracted the temperatures in their cores rose higher and higher... and then they contracted further, and helium began to react and produce the other elements that are fairly abundant,

carbon, nitrogen, oxygen.

At certain moments during this process the stars become unstable and then they spew out material. This happens in great explosions. In Supernova explosions"

So, the first stars created the new elements like carbon and oxygen as they got hotter and denser. And when they exploded at the end of their lives after billions of years spewed these atoms into the universe. This allowed the new atoms to become part of newly born stars and create even more elements which then spewed out into the universe.

It's the life and death of stars that have created the 92 elements that make up everything. This is why scientists are fond of saying that we are all made from stars. And when we cease to be our atoms will be recycled and will construct something else in the universe since atoms are indestructible.

However, all of this relies on some preconditions for which scientists have no valid answer.

The preconditions are the values of the laws of nature, such as the strength of gravity, the speed of light, the charge of electrons, the value of the Cosmological Constant and many others. The values for all these laws were set by default automatically at the start the big bang.

And these values are finely calibrated to an unimaginable degree.

And it is this careful calibration which has allowed

everything to come into being and to exist together in almost balletic harmony.

If for example, the value of gravity was even slightly different to what it is then stars would never have formed. No other elements would have been created. And the universe would have remained just a cloud of dust.

Dust to dust. Ashes to ashes.

But why are there these laws of nature? And why are they so calibrated, so finely tuned to allow the universe to come into being and for us to exist? How can it be that the laws of nature are just so right?

It didn't make sense. The default values were too precise.

Scientists started scrambling around to find a solution. And when they couldn't find one, they decided to take a leap of faith, as big as any person of religion, in a theory that should be scientific heresy.

Chapter 2

For millennia, since the time of the Greeks, who first started thinking about these things in any meaningful way, it was believed that the planet Earth was at the centre of the universe. And everything we could see, the moon, the other planets in our solar system, and even the sun, revolved around the Earth.

The Earth was the most important thing in the universe. It remained stationary while everything else revolved around it. This was known as the Geocentric view.

The Earth was special. It had meaning.

But Scientific discoveries upended that notion.

In the 16th century Copernicus was the first suggest that actually the Earth was not special. It was just a planet that orbited the sun, as did all the other planets in our solar system. And it was the sun that was at the centre of the solar system.

Dennis Danielson – Editor the Book of the Cosmos

"The idea of the moving Earth seemed to violate some fundamental principle. But Copernicus somehow had the mental power to imagine what even to him seemed absurd. So, he thought the impossible; the Earth moves. And once you imagine the Earth moving instead of the sun, the mathematics of that cosmic machine started to make sense.

It was the key that unlocked one of the mysteries of the universe"

It seemed that the Earth occupied no special place in the universe.

Max Tegmark – Professor MIT

"We realised that we are not the centre of the universe. We're just living on a spinning ball; one planet among many, in a solar system among many, in a galaxy among many."

Bit by bit the Earth and our place in the universe went through a series of demotions. Neither the Earth nor human beings were special.

Scientists even had a name for this view. It was called "The Principle of Mediocrity."

Jay Richards – Philosopher and author

"This principle says that our location and our status are mediocre, they're unexceptional. As a result, we should not assume that we are in anyway privileged. Or that the universe was designed with us or beings like us in mind."

Scientists came to the conclusion that the Earth is unexceptional and has no particular meaning and human beings are not special in anyway. We just happened to get lucky, and our existence has no particular significance.

In the 1970's a space craft sent out by NASA that had completed its mission to explore the outer planets was sent instructions by the famous astronomer Carl Sagan, to turn the camera around and take a picture of the

Earth from the edge of the solar system.

The picture was quite revealing. It showed the Earth as an insignificant, tiny speck of dust bathed in a beam of light.

Carl Sagan – Astronomer – writer – broadcaster

"Because of the reflection of sunlight, the Earth seems to be sitting in a beam of light, as if there were some special significance to this small world. But it's just an accident of geometry and optics. Look again at that dot. That's here, that's home, that's us. Our posturings, our imagined self-importance. The delusion that we have some privileged position in the universe are challenged by this pale light. Our planet is a lonely speck in the great enveloping cosmic dark."

Poetic it may be but that sounds pretty depressing.

It was all just by chance. The Earth just got lucky to have an environment capable of supporting life. And we got lucky enough to become the dominant life form on the Earth.

Yes, sure the odds were high. But someone, somewhere wins the lottery. We just happened to win the lottery. We and the Earth and the solar system and the universe are just a statistical fluke.

But it may not be like that at all.

In the last hundred years scientists started to discover the exact values of the laws of nature. All of which had to be at a certain precision otherwise the universe just would not exist.

And the more values and constants of nature they discovered, the more it started to look like the universe had a structure and design, that it could not have possibly come about by chance.

This was pretty disconcerting. Scientists had thought that they had things figured out but now had to unpack everything and look again. Maybe things were not as simple as they seemed.

Chapter 3

Scientists had worked out that the first stars were made of just hydrogen and helium and then these stars made other elements like carbon and oxygen. This was rather crucial since we human beings are made of carbon and need oxygen to breath.

But what processes and laws made these new elements like carbon and oxygen in the heart of stars?

Scientists in the 1950s started looking at how these elements were created and came to the conclusion that the process by which the helium nuclei combined to make carbon seemed to be by design and not chance.

It was the same with the creation of the oxygen atom. The process to create oxygen was incredibly fortunate as well.

Fred Hoyle - Astronomer

"I do not believe that any physicist who examined the evidence could fail to draw the inference that the laws of nuclear physics have been deliberately designed with regard to the consequences they produce inside stars."

The word "deliberately" must have made scientists start to think.

Eager beavers they are, scientists started to look at some of the other laws of nature like:

Gravitational Constant (G): The strength of gravity,

which determines the attractive force between all masses.

Speed of Light (c): The maximum speed at which information and matter can travel in the universe.

Fine-Structure Constant (α): A dimensionless constant that characterizes the strength of electromagnetic interactions.

Mass of Elementary Particles: The masses of fundamental particles, such as electrons, protons, and neutrons, play a role in the formation of atoms and molecules, which are crucial for the chemistry of life.

Strong Nuclear Force: The strength of the strong nuclear force, which holds atomic nuclei together.

Weak Nuclear Force: The strength of the weak nuclear force, responsible for processes like beta decay in stars.

Cosmological Constant (Λ): A parameter related to the energy density of empty space and the accelerated expansion of the universe.

Electromagnetic Fine-Tuning: The balance between electromagnetic forces, such as the electromagnetic repulsion between protons in atomic nuclei, which prevents them from collapsing.

Nuclear Resonances: The existence of specific energy levels in atomic nuclei, known as resonances, that enable the production of elements in stars through nuclear fusion.

The Mass-Energy Equivalence ($E=mc^2$): The famous equation developed by Einstein that relates mass and energy, playing a crucial role in nuclear reactions and the energy output of stars.

Neutrino Mass: The tiny mass of neutrinos, which influences processes in the early universe and the dynamics of stars.

And the list goes on.

All of these laws and many more were finely tuned to not only enable the universe to come in to being and continue to exist but also allow life to come into existence.

If any of the laws were slightly different everything would pretty much fall apart. And there were no known reasons why the values of these laws should be set as they are.

Martin Rees - the Astronomer Royal

"It looks in some respects as though our universe is rather special. We know the universe allowed our emergence but it's quite easy to imagine a universe with slightly different properties in which neither we nor anything as complicated as us could exist.

We can imagine as it were, turning the knobs which were set up at the time of the big bang to determine how it expanded and what it was made of. And if we turned the knobs very slightly, we find that we would end up with a universe that would not be so propitious for the emergence of life."

Physicist and writer Paul Davies concurs:

"The really amazing thing is not that life on Earth is

balanced on a knife edge, but that the entire universe is balanced on a knife edge. You see even if you dismiss mankind as just a mere hiccup in the great scheme of things there's still the fact that the entire universe seems unreasonably suited to the existence of life, almost contrived. We might say a put-up job."

The underlying nuts and bolts of the universe, the forces that hold it all together, the laws and constants of nature, interlock precisely and allows the universe and us to exist. How can this be?

Dennis W. Sciama- Physicist

"It turns out that if you change just a little bit the laws of nature or you change a little bit the constants of nature, like the charge on the electron, then the way the universe develops is so changed that it's very likely that intelligent life would not be able to develop."

More and more scientists were coming to the same conclusion.

Dr. David Deutsch – Oxford University

"If we nudge one of these constants just a few percent in one direction, then stars burn out within a million years of their formation. No time for evolution. And if we nudge it just a few percent in the other direction then no elements heavier then helium form, so no carbon. No life. Not even any chemistry, no complexity at all. And if we could alter the relative masses of two atomic particles, the proton and the neutron, by just a fraction of a percent, atoms would

be unstable. There'd be no stars, no light, no warmth, no structure at all; just chaos."

One of the discoverers of the cosmic microwave background put it rather succinctly.

Arno Allan Penzias - Nobel laureate in physics

"Astronomy leads us to a unique event, a universe which was created out of nothing, and delicately balanced to provide exactly the conditions required to support life."

Despite the growing evidence of a fine-tuned universe set precisely for everything to come into being and exist, scientists for a while were confident of being able deny the possibility that there was a purpose to all that we see.

It was all just coincidence or that at some point a theory would come along and explain all the fine tuning.

Leonard Susskind – Theoretical physicist

"The general view of this for most physicists is that these fine tunings are largely accidental. That the constants of nature are determined by mathematical principles which have nothing whatever to do with our existence. Impersonal, mathematical and we were just incredibly lucky that that mathematics happen to give rise to a universe with all this kind of fine tuning just precisely so."

But the fine-tuning conundrum still bothered scientists. They just could not wish it away. And it got worse. A new anti-gravity force was discovered called the Cosmological Constant. And when scientists calculated how it affected the evolution of the universe, they found it had to be fined tuned to a level unheard of in any other force of nature.

Leonard Susskind – Theoretical physicist

"The fine tunings, how fine-tuned are they? Most of them are one percent sort of things, in other words if a thing is one percent different everything gets bad. And the physicists could say maybe those are just luck.

One the other hand this cosmological constant is fine tuned to one part in ten to the one hundred and twenty. A hundred and twenty decimal places. Nobody thinks that's accidental. That is not a reasonable idea. That something is tuned to a hundred and twenty decimal places just by accident."

The cosmological constant needs to be set to one part in a trillion, trillion, trillion, trillion, trillion, trillion, trillion, trillion, trillion, trillion. Otherwise, the universe would have been so drastically different that it would have been impossible for human being to emerge.

Scientists who believed in a random chance universe were caught between a rock and a hard place. On the one hand they could not entertain the merest possibility of a created universe, but on the other could not accept that a universe just happened to have exactly the constants and laws of nature required for us to exist.

They tried but just could not wish away the fine-tuning. The laws worked too well. There was too much synchronicity in the cosmic dance. It was too beautifully choreographed.

It was like watching the Bolshoi Ballet performing a perfect rendition of Swan Lake. But all the dancers had come together by accident on the night, complete strangers to each other. No one had chosen them, no one had called them. They just turned up on the night. Every dancer, male and female was exactly the right weight, height, size, sex and had perfect agility and dancing skills required for the role they performed in the ballet. The composition of which they could not possibly have known about in advance.

There had to be a reason for this beautiful choreography.

And it could not have anything to do with a creator.

And so, they came up with a theory that accepted a designed universe but crucially without an actual designer.

They called the theory The Anthropic Principle.

Chapter 4

The physicists who believe in the Anthropic Principle are an interesting bunch. And the theory itself is rather interesting too. It is almost spiritual in its way. It goes so far as to say that there is a purpose to the universe and a reason for our existence.

The fine tuning is by design. But it has nothing to do with a creator.

The Anthropic Principle is much misunderstood. Many people perhaps don't appreciate that it is not even a single hypothesis. There are various elements to the principle.

The first and most simple element is the Weak Anthropic Principle.

This principle counters the Principle of Mediocrity and the Copernican Principle which posits that we human beings are just an accident of nature. That we are nothing special, we have no reason or purpose for being in the universe.

Brandon Carter – Formulator of The Anthropic Principle theory

"...the Copernicus Principle which states that our own situation in the universe is in no way privileged is just not true. The very fact that we are here places restrictions on what the universe can look like."

Basically, we are kind of special because if the universe

was any other way, we would not exist to observe the universe in the form it is.

Frank Tipler - physicist and cosmologist and writer

"So far there is no argument. In the form Carter originally put it The Weak Anthropic Principle merely says that we are observing the universe from the viewpoint of a very particular species, Man."

So, this version just states that the universe's parameters and conditions may appear fine-tuned for life because we can only exist in a universe that is suitable for our existence. If it were not so, then the universe would not have observers who could make that observation.

This is really just a statement of fact. We exist because the parameters of the universe are set to allow us to exist.

But this does not really address the fine-tuning. Why are the parameters of the universe set to allow us to exist. Brandon Carter came up with the second element of the theory; The Strong Anthropic Principle.

Brandon Carter – Formulator of The Anthropic Principle theory

"To cope with this kind of puzzle I formulated the Strong Anthropic Principle. This says our universe MUST have those properties that allow intelligent life to develop in it at some stage."

This really put the cat amongst the pigeons in the scientific world.

Martin Rees - the Astronomer Royal–

"Must?"

Dennis W. Sciama- Physicist

"Must?"

Michael Redhead – Philosopher of physics

"Must?"

Scientist now started to splinter as to the validity of this second element of the Anthropic Principle.

Frank Tipler - physicist and cosmologist and writer

"Must have? This really brings the disagreement to the fore because in that little word "must" is concealed a metaphysical presupposition that is really anathema to science."

This second element of the principle states that the universe had to have the properties that allow for the existence of intelligent life. The universe was compelled to give rise to not only conscious beings but conscious beings capable of uncovering and understanding the laws of nature.

The universe it seems brought us into existence so it could know itself. An interesting and very philosophical view but not very scientific. Not surprisingly some scientists disagreed with this element.

Michael Redhead – Philosopher of physics

"In order to predict the values of the fundamental constants of nature, you don't have to use the existence of

Man, we could invoke a crocodile principle if you like. From the existence of crocodiles or dragonflies or any other form of life, we could use that to predict what the constants are."

The argument against was that our world had life forms other than just human beings, so it does not mean human beings had to come into existence.

But the believers in the Strong Anthropic Principle felt that that was missing the point.

Dr. David Deutsch – Oxford University

"Yes of course, all this special features of the universe gave rise to the crocodile...but we have something the crocodile does not have. We understand the laws of physics. And it is understanding that the Anthropic Principle makes into a key feature of the universe."

Crocodiles and dragonflies and other life forms are just creatures of instinct. They are not able to query and find answers on the workings of the cosmic ballet. But we human beings acting as observers are able to make sense of the universe. As such without us the universe has no meaning.

Dennis W. Sciama- Physicist

"It's a puzzling and I think rather non-trivial thing that we can discover so much about the universe. We are after all a part of the universe, so we have a system which is developing according to the laws of nature.

Localised complex systems like human beings developed,

with brains; and lo and behold using those brains and using physical systems which obey the laws of nature, to investigate nature, we can find out what those laws are. That's an amazing thing. And I think the full significance of the fact that the structure of the world permits that to happen is not at all clear."

Dr. David Deutsch – Oxford University puts it a bit more simply.

"Why should the laws of the universe be understandable? By us! Well, either it's a fantastic coincidence or there's some deep reason why it had to be that way. And if it had to be that way, then we occupy a very special place in the universe indeed."

Human beings after all were not mediocre.

John Archibald Wheeler - Theoretical physicist

"An old legend describes a dialogue between Abraham and Jehovah. Jehovah chides Abraham "you would not even exist if it were not for me."

"That I know, oh Lord" replies Abraham. "But you would not be known if it were not for me.""

Max Tegmark – Professor MIT

"If you look at the cosmos, sure we're small but suppose there were no life in the universe, wouldn't all this beautiful stuff out there be a complete waste if there were no one to behold it? I think so. I think it's only life which

gives any sort of meaning to the universe. Particularly if it turns out that we are the only life in the observable universe."

And so, the Strong Anthropic Principle gained a lot of traction.

But some scientists were not content with the Strong Anthropic Principle. They went even further. Drawing their inspiration from the weird and mind-bogglingly complex world of quantum mechanics, they came up with a new element to the Anthropic Principle. They called this the Participatory Anthropic Principle.

John Archibald Wheeler - Theoretical physicist

"The Strong Anthropic Principle states that the universe must give rise to life at some point in its course. Modern quantum theory, the overarching principle of twentieth century, leads to quite a different view of reality. The view that man, or intelligent life or communicating observer participators, are the whole means by which the universe is created. Without them, nothing."

This was quite mind blowing! It is our observations that brings things into existence. We are in fact the creators of the universe.

Dr. David Deutsch – Oxford University

"...there is something very attractive about it. If it were true, we wouldn't need any other explanation for the whole creation of the universe. It would be observers that bring the universe into existence, not just in the present but throughout the whole past right back to the big bang."

The act of observing brings the cosmos into being. This is possible for the past too since when we look at space we are looking back in time. The further back we are able to observe using our space telescopes the more pieces of the universal puzzle we are simultaneously creating and putting into place.

Needless to say, the Anthropic Principle in all its elements was extremely unsatisfactory to a lot of scientists. Some clearly thought it was bonkers, especially the Participatory element.

But it was all they had. They saw it more of a stopgap then anything they could take seriously. It had little scientific merit.

Martin Rees - the Astronomer Royal

"The question of the status of the Anthropic Principle is whether it is just a stopgap way of satisfying our curiosity until we have a proper physical theory or whether in a sense it can be formalised, so it is a physical theory. Everyone would agree it is at least a stopgap."

There are still many adherents of the Anthropic Principle, especially the strong element, but most saw it as a stopgap. It was not a theory that properly explained the fine-tuning. And the participatory element was just too out there. And worst of all it came way too close to be having a spiritual if not a religious element.

No, the Anthropic Principle was attractive but did not have sufficient substance.

There had to be another answer. One that did not

require human beings to have any special purpose and one that would explain the fine-tuning that was causing scientists to veer dangerously to a spiritual mindset.

Chapter 5

Scientists who believed in a random chance creation of the universe were in a bind. Their own discoveries were pointing them towards an intelligent designer. A universe especially created for us to exist. But this clearly went against the scientific holy scriptures.

Leonard Susskind – Theoretical physicist

"Physicists dislike mixing religion with physics. I think they were somewhat afraid if it was admitted that the reason the world is the way it is has to do with our own existence. Then that could be hijacked by the creationists, the intelligent designers and of course what they would say is "yes, we always told you so. There is a benevolent somebody way up high in the universe who created the universe exactly so that we could live." I think physicists shrank at the idea of getting involved in such things."

So, like flat Earthers who discover evidence that the Earth is actually round and not flat as they believed, they started scrambling around for theories that would explain away the evidence they had uncovered. The belief in a flat Earth was unshakeable and any evidence found that contradicted it had to have some explanation. It could not be that there was a possibility that they were wrong, and the Earth was round. That was not something even worth contemplating.

Unfortunately, The Anthropic Principle theory just wasn't cutting it. It was too philosophical and had little or no scientific basis anyway.

No, there had to be something else. Something that would explain all the fine tuning that went to the level of chance that was trillions and trillions to one. And whatever they came up with could have nothing to do with a creator.

And lo and behold, scientists found the perfect solution.

Martin Rees himself was one of the proponents of a much more elegant theory. A theory that solved the fine-tuning conundrum at a stroke. And it made perfect logical sense.

It was really quite beautiful. And exactly what was required.

And the name of the theory? The Multiverse.

What if our universe, with its very specific laws of nature that allowed for all the fine-tuning, was just but one of multiple universes? There may be trillions and trillions, trillions and trillions, trillions and trillions, trillions and trillions, trillions, and trillions of universes out there. Each of them with their own specific constants and laws of nature different to ours. Then it is not inconceivable that our universe just happened to have the right conditions for life to emerge.

If there were many trillions of universes with their own big bangs that set initial specific conditions, then surely one of those big bangs was likely to have those conditions set precisely to allow for the emergence of life.

Each week someone wins the lottery. We just happened to win the lottery.

Martin Rees - the Astronomer Royal

"If there be many big bangs and if, and this is a second assumption, the outcome of those big bangs were universes governed by different physical laws, then we could imagine that there would be one universe governed by any particular law we care to envisage. Therefore, it would not be at all surprising if there should be one universe that was tuned."

It was a brilliant solution.

A famous analogy of it is if you go into a large clothing warehouse, and in this huge, cavernous warehouse there is just one rail, and on that rail, there is just one suit. And that suit was perfect for you; just the right length trousers and jacket, just the right waist size, the perfect colour and style you like, and it fitted you extremely comfortably. Then you'd probably start to look around to see who put it there. Someone clearly had you in mind hanging that suit up there. Otherwise, how could it fit you so perfectly.

But if you went into a clothing warehouse with hundreds of rails and thousands of suits, all of different colours and sizes and styles. Then you would not be surprised to find one that was exactly right for you.

The concept of the Multiverse was just what scientists

were looking for. Fine-tuning explained. No need to consider the possibility of a specially created universe.

Leonard Susskind – Theoretical physicist

"We now have a natural mechanism to explain why there is all this diversity out there, which in turn eliminates the need for the fine-tuning that some people might have liked because they could say there is a fine-tuner. We don't need a fine-tuner."

Scientists breathed a huge sigh of relief. If they were Flat Earthers then they had found a concept which allowed theoretically at least for the Earth to be flat, and they could now explain away the evidence of the curvature they had discovered.

The Earth was flat, exactly as they had said all along.

The theory was logical, it made mathematical sense, and it did away with all the fine-tuning. Now there was a reason why we found ourselves in a universe that had all these laws and constants set so very precisely. And it did not have anything to do with a creator.

They could continue to believe in the scientific holy scriptures.

Despite having pretty much no evidence for it, the multiverse theory has taken a life of its own and has been expanded to various types and levels.

One concept has four levels:

Level I: An extension of our universe

Level II: Universes with different physical constants

Level III: Many-worlds interpretation of quantum mechanics

Level IV: Ultimate ensemble

Another concept has nine different types with names like Quilted, Landscape, Quantum, Cyclic and so on.

This is almost reminiscent of QAnon followers who try to decipher the meaning of the latest Q drops on what is really going in the world that the "Deep State" don't want you to know about.

Scientists were extremely excited about this new theory they had formulated.

However, things were not all rosy in the multiverse world.

There were two very fundamental problems with the theory.

One was that it still did not explain what made all these many trillions of universes with their own specific set of laws come in to being. If anything, it made the problem worse. Before they had to explain how one big bang happened. Now an explanation was needed for many trillions of big bangs.

Cause and effect. The scientific principle of cause and effect just could not be got round. Scientists can delay it with concepts of previous big bangs, starting of big new bangs and such. But that just leads to an infinite regress. At some point you must have cause and effect.

The Multiverse theory, as beautiful as it was, did not address cause and effect.

Generally, science has not attempted to answer this question. There is an argument to say this is not a scientific question but a philosophical one and perhaps best left to philosophers. This seems very unsatisfactory. If scientists do not even want to entertain the possibility of a created universe, then surely, they have to have some explanation for an alternative answer?

The second fundamental problem was a big hurdle to cross too. What evidence is there of a multiverse? Science is all about evidence, data, and proof. There was nothing of any scientific note to back up this theory.

Not only that but how can this theory be even tested? No telescopes will ever be able to peer at another universe. No scientific instruments can ever be created to measure another universe. How is this even a scientific theory?

It is not.

It is a matter of faith.

If you want to do away with all the anomalies scientific discoveries have uncovered such as inflation, and dark matter and dark energy and also do away with all the fine-tuning by believing in this theory in order to keep an atheistic outlook, then that is perfectly fine. We are all entitled to our beliefs.

But it must be accepted that believing in the multiverse to do away with all the evidence that clearly indicates that there is at the very least a possibility that the universe was created and that we exist for a reason, is a matter of faith and as such perhaps Atheism should be considered to be a religion in a way.

And to be fair there is a big debate in the scientific world as to the validity of the concept of the multiverse. Not all scientists believe in this theory, and they also point out that if you believe in a multitude of universes then you are no different from someone who believes in God.

Paul Davies – Physicist and writer

"...how is the existence of the other universes to be tested? To be sure, all cosmologists accept that there are some regions of the universe that lie beyond the reach of our telescopes, but somewhere on the slippery slope between that and the idea that there is an infinite number of universes, credibility reaches a limit. As one slips down that slope, more and more must be accepted on faith, and less and less is open to scientific verification. Extreme multiverse explanations are therefore reminiscent of theological discussions. Indeed, invoking an infinity of unseen universes to explain the unusual features of the one we do see is just as ad hoc as invoking an unseen Creator. The multiverse theory may be dressed up in scientific language, but in essence, it requires the same leap of faith."

Beautifully put.

ADAM SUSSKIND

Closing argument

This book has not been written to convert atheists into believers. Nor does it profess to prove the existence of a creator. What I hope has been highlighted is that scientific discoveries made over the last hundred years or so clearly show that there is a very real possibility that the universe was created.

Science and religion can live together and are not mutually exclusive.

In previous ages when scientists looked for answers on the workings of the universe, they did so to seek to understand the mind of God. Famous names like Newton, Darwin, Copernicus were men who believed in a created universe.

Perhaps the one theory that has had so much impact on turning scientists to atheism is Darwin's theory of evolution. There is no question that Darwin's theory has a great deal of validity and species have evolved and changed over time, that new species come from pre-existing species, and that species share common ancestors.

But the theory of evolution is not necessarily complete.

The two greatest theories of the twentieth century, Einstein's theory of relativity and the theory of quantum mechanics are both valid and explain so much, but they are incomplete. They have gaps. They don't work with each other. Clearly there is something missing. Scientists are still trying to formulate the "Quantum Gravity" theory that will explain everything.

Darwin's theory also has gaps.

It is clear that of all the hundreds of millions of species that have emerged on planet Earth over billions of years none have been like human beings.

All other species act on instinct and survival. They evolve to suit their environment. They live and perish on how well they adapt to changing conditions. Dinosaurs were an extremely successful species for example. They ruled planet Earth for around 165 million years. That's a very long time. But what did they achieve in all that time, apart from surviving so long?

And you can say that about every other species who came and went or still survives on Earth.

We human beings are different though. It is not just about survival for us.

We can think.

"I think therefore I am."

We have consciousness. There is no evolutionary reason why we should have consciousness. We contemplate our very existence. We search for answers to how we were created. Why do we need to do that from an evolutionary perspective?

We human beings have been on this planet for a fraction of the time that Dinosaurs ruled Earth. But we have taken over the planet like no others before us.

Our acceleration as an intelligent, sentient species that can contemplate its very existence has been phenomenal. It's almost as if someone designed us to be this way.

Designed us to be able to think. To make moral decisions. To try and figure out right from wrong. To believe or not to believe in a creator.

This is where the theory of evolution falls short. The theory is not complete.

But some scientists don't want to think about such things. The belief is that science can explain everything. And it's possible that at some time in the future, science will find answers to the gaps in the cosmological bible. We may be able to uncover answers to how inflation happened and why it stopped at just the right time, or how stars are sticking together despite there not being sufficient visible matter. Scientist may even discover what the so-called Dark Energy is all about. But that may be uncovering how a creator put the universe together. It won't be able to explain how or why it was put together like that.

Science has no answer to those questions. And it doesn't look like it ever will.

The scientific world has taken a sharp turn in recent decades in the direction of Atheism. And believe that there is no need to invoke a God as a creator of the heavens.

In fact, people who believe in a god who created the universe are viewed as naive at best or deluded fools at worst. They are looked down upon in a patronising way and are referred to disparagingly as creationists.

Look at them, the poor deluded fools. They genuinely believe a god created everything.

But there is a very real possibility that it is the scientists themselves, who refuse to accept the evidence staring them in the face that the universe was quite possibly created, who are deluding themselves.

Like the flat Earthers who, instead of considering the evidence they have uncovered and accepting that there is at least the possibility that the Earth is round, scientists are sticking to their preconceived notions of a universe created just by chance.

And they come up with more and more ingenious and wildly speculative theories, that cannot be proven or disproven, to continue to keep their faith.

And that is precisely what it is. Faith. And they are entitled to it. But they must accept that they are just as much religious as any believer in a traditional religion.

And that their religion can be classed as Atheism.

If you have enjoyed reading this book or found it interesting and or illuminating, please consider leaving a positive review. It would be highly appreciated.

SOURCES

The majority of the quotes used in this book were sourced from the following television documentaries:

"Dancing in the Dark - The End of Physics?" Narrated by David Mitchell (BBC Horizon)

"What We Still Don't Know" – Hosted by Astronomer Royal Martin Rees:

Episode 1 – Are We Alone?
Episode 2 – Why Are We Here?
Episode 3 – Are We Real?

"The Mystery of Dark Energy" (BBC Horizon)

"The Privileged Planet" – Narrated by John Rhys-Davies

"Is Everything We Know About The Universe Wrong?" (BBC Horizon)

"THE UNIVERSE - Out of Nothing: Infinity" WELT

"The Anthropic Principle" (BBC Horizon)

"What Happened Before The Big Bang" (BBC Horizon)

The quote from Neil DeGrasse Tyson was found using Google Search.

The quote from Paul Davies in the final chapter is from Wikipedia.